The Rational Scientist

July - August - September
Copyright Grinning Monkey Publishing
All Rights reserved

ISBN-13:978-1721778843
ISBN-10:1721778845

THE RATIONAL SCIENTIST

A quarterly Magazine

Available on therationalscientist.com in E format for immediate download, and on Amazon in 8.5 x 11 paper format.

Our mission is to introduce you to the Rational Scientific Method and to expose current mainstream nonsense such as Relativity, Quantum Mechanics, String Theory, Big Bang, Black Holes, Faster Than Light, Warped Space, Multi-Dimensions and TimeTravel.

Science magazines have gone astray. Today, the popular rags of science are more interested in scientific fantasy, and political debate or social justice than they are in explaining phenomena using objects. But, then, mainstream science abandoned using objects to mediate phenomena long ago in lieu of abstract theoretical mathematical descriptions and reification.

This issue is compiled from the first in a series of books entitled "Rope Hypothesis and Thread Theory." Light, Gravity, Electricity and Magnetism are all mediated by one unifying physical mechanism comprised of Thread.

Get the paperback, E-book, or audio book version of the Rope Hypothesis and Thread Theory on Amazon.

We hope you have at least some degree of familiarity with the Rational Scientific Method and the Rope Hypothesis. If not, then I highly suggest that you:

Join the Rational Scientific Method and Rope Hypothesis facebook groups.

Find the books on Amazon, find Monk E. Mind on facebook and find out more about TRS on the website: www.therationalscientist.com

Editor and Chief
Monk E. Mind

Executive Editor
Monk E. Mind

Creative Director
Monk E. Mind

Contributors
Monk E. Mind
Daniel Ferguson

Artwork
Monk E. Mind
Daniel Ferguson

The Rational Scientist
Published By

GRINNING MONKEY PUBLISHING
monkemind@gmail.com

COVER ART BY
Monk E. Mind
and Daniel Ferguson

In this special issue, sprinkled throughout, is the artwork of Daniel Ferguson, rational scientist and supreme artist. As we apply the rope model towards a theory of Threads for various aspects of physics, Daniel sketches out his ideas. In this way, it is always clear what he is proposing. Daniel understands, as do all rational scientists, that Physics requires objects, and the language of science is illustration.

CONTENTS

JULY
AUGUST
SEPTEMBER
2018

07 GRAVITY BASICS

10 LIGHT GRAVITY AND MAGNETIC MOMENT

12 GRAVITATION AND ELECTROSTATICS

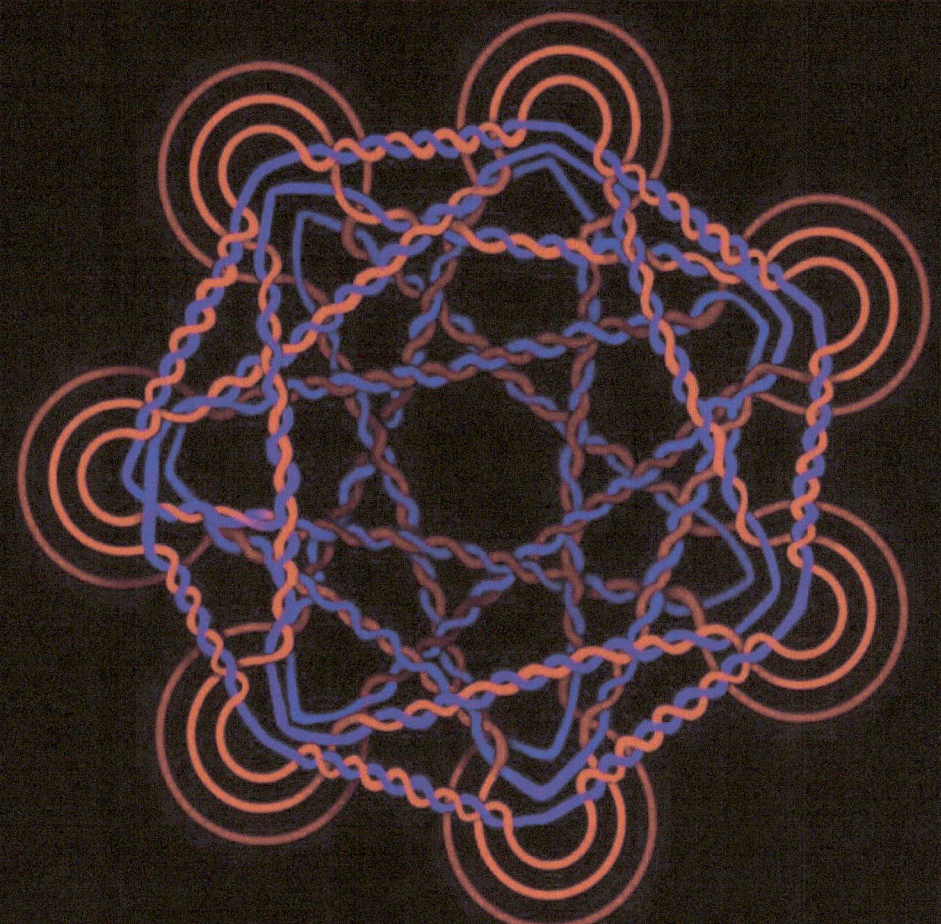

17 PHOTOEECTRIC EFFECT AND THREAD THEORY

20 LIGHT AND SOUND HOW ARE THEY DIFFERENT?

21 LIGHT AND SOUND HOW ARE THEY ALIKE?

Page 26 Refraction Reflection and Diffraction   **Pages 38-42** Advertisements
Page 32 Static Electricity and Thread Theory   for Member's Products

From the Editor
MonkEmind

Bill Gaede is the originator of the Rope Hypothesis (RH). Being the genuine scientist and man of integrity that he is, he openly requests that persons discuss, argue and debate the hypothesis applying the rope model towards a Theory of Threads for all scientific topics from the atom to "The Universe." As such, we have given our best effort in bringing forth the Rope Hypothesis to the public eye. Whether novice, science buff, student or professor much can be gained by exploring the simple architecture of the rope model, faithfully applying the Rational Scientific Method (RSM), and building a grand unified theory upon its foundation.

This meager attempt is in no way representative of "settled" science. For if there is anything that we have learned, it is that there is NO settled science. In the spirit of REAL science, that is, explanations for reality, there is only ever possible and NOT possible. You, dear reader, must decide this for yourself. We invite you to come by the RSM and RH facebook groups, to contact me or Bill and other rational scientists, and to discuss vigorously all topics big or small. We welcome your discourse and your dissent.

Now join me as I apply the Rope Hypothesis towards a Theory of threads for electricity, magnetism, gravity and light. In this offering I attempt to show the interconnectedness of light, gravity, electricity and magnetism via the simple architecture of "the rope." We only scratch the surface due to the limited space, the depth of Rope Hypothesis, and the unlimited application of Threads.

The Rope Hypothesis

A single continuous thread weaves in and out of all objects in the universe forming a rope like structure with "knots" or atoms. There are no discrete particles anywhere.

The ropes are comprised of an Electric and a Magnetic thread. The two (E & M) threads are anti-parallel strands that twist around each other like a DNA helix and separate at atoms. The M thread wraps around the outside of the atom and the E thread goes to the center of the atom. Atoms expand and contract and the threads unravel, or unwind, reel in and reel out. The pumping action of the atom is "felt as a torsion signal along the rope." This is light. The frequency is the number of lengths per unit length. The length of the links changes in atoms of different medium. Amplitude is related to link height.

Rope Hypothesis provides the physical mechanism by which Thread Theory explains gravity and Immediate Action at a Distance (IAAAD).

The tension between objects is the net result of all ropes. This is the force we call gravity (pull).

"As one object approaches another, the EM ropes fan out as a function of decreasing distance and cause the acceleration of one to the other." – Bill Gaede

Thread Theory explains light:

It explains reflection, refraction, diffraction, wave/particle duality, wavicles, gravitational lensing, polarization, slit experiments and Olber's Paradox; all atoms are interconnected by a physical medium which mediates light.

TT explains how it is that electricity and magnetism run perpendicular to each other; Anti-parallel electric and magnetic threads wrap helically around each other mediating torsion and tension.

TT explains electric current; enmeshed E shells form electron serpentines, spinning together to create what is called electric current. No holes moving left or beads moving right.

TT explains how magnets attract and repel; Sweeping magnetic threads either attract or repel each other similar to jumping ropes interacting.

TT explains why galaxies rotate at about the same rate on the edge as at the center; all atoms are connected by EM ropes. There is no mysterious Dark Matter or energy.

TT explains covalent bonding; M threads from adjacent atoms spin in the same direction drawing atoms in. E shells merge, there are no discrete electron beads exchanged between atoms.

TT explains ionization; Expanded E shells; no discrete electron particles or electron clouds or orbitals are involved.

TT explains Beta decay; Atoms "pick up" and "drop" crisscrossing ropes called neutrons.

TT explains Ray Reversibility; Interconnecting ropes mediate light in both directions simultaneously. Explains Einstein's light/train Gedanken, and how retroreflectors on the moon "bounce back" the light as the earth and the moon are moving.

TT explains Newton's Laws of motion; "It is inconceivable that inanimate Matter should, without the Mediation of something else, which is not material, operate upon, and affect other matter without mutual Contact…That Gravity should be innate, inherent and essential to Matter, so that one body may act upon another at a distance thro' a Vacuum, without the Mediation of any thing else, by and through which their Action and Force may be conveyed from one to another, is to me so great an Absurdity that I believe no Man who has in philosophical Matters a competent Faculty of thinking can ever fall into it." – Newton

It's all about interconnecting ropes!

TT explains Maxwell's equation; $f=c\lambda$

It is the equation of a rope!

TT explains Mach's Principle and what he meant by; "[The] investigator must feel the need of... knowledge of the immediate connections, say, of the masses of the universe. There will hover before him as an ideal insight into the principles of the whole matter, from which accelerated and inertial motions will result in the same way." - Mach

Those connections are the interconnecting ropes of the Rope Hypothesis.

TT explains why both gravity and electrostatic "force" are stronger or weaker based on "the inverse of the distance squared."

$F_g = G.M_1.M_2/D^2$
$F_e = K.Q_1.Q_2/D^2$

Thread Theory explains much more!

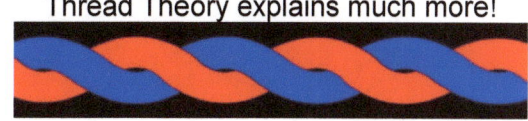

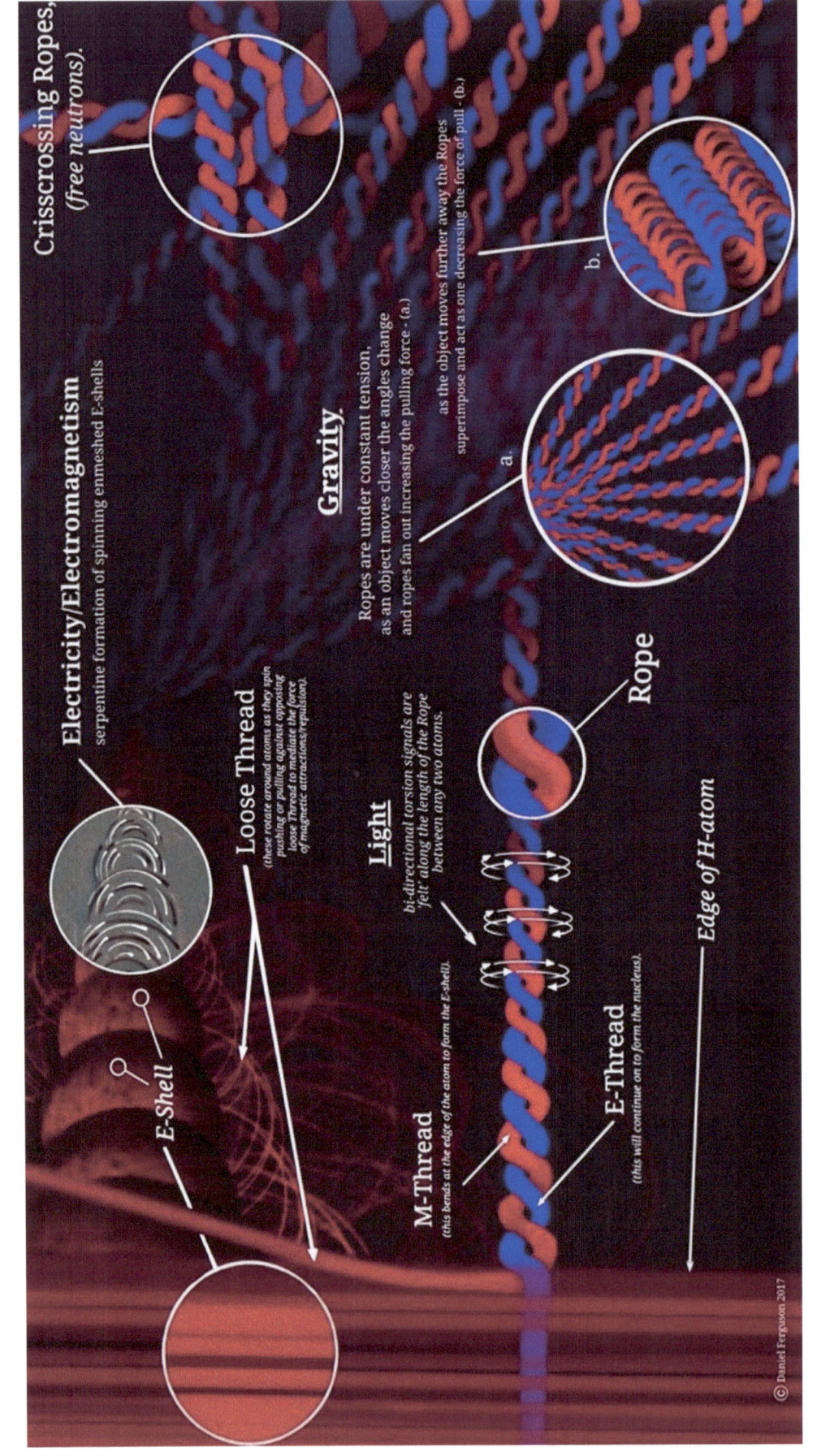

Gravity Basics

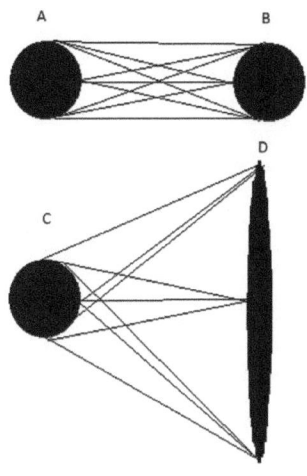

Objects A,B,C, and D are all made from exactly 1 litre of lead. Objects A, B and C are balls formed in the same mold. Object D is a flat disc. Should the strength of gravity between objects A and B be significantly different from the strength of gravity between objects C and D?

For gravity, we have tension along ropes between the two objects. When objects are closer together the ropes fan out and the net pull is greater.

Mich: In this diagram the objects are not brought closer together, one of them is merely flatter and longer than the other. However, the ropes fan out, rather drastically, in this scenario as well. So should the gravitation between C and D be greater than the gravitation between A and B? I don't think it should, but I can't understand in the context of RH why it wouldn't."

Assuming the same number of atoms in all 4 objects, the actual pull (tension) is the same between A&B and C&D. Since D is wider the net pull is greater because the ropes are fanned out and the objects will move towards each other faster.

Mich: "So a flat plate will fall faster than a round ball. I think we should pause and think more on this."

In a vacuum they fall at the same rate. If there is air resistance, then, no.

Mich: "We'll assume vacuum."Since D is wider the net pull is greater because the ropes are fanned out and the objects will move towards each other faster." So changing the shape of an object would alter the velocity of its fall towards the other object? Imagine then that objects A and C are moons and that objects B and D are satellites. It would seem that a broader satellite would fall more quickly than a narrower one."

No, it's the location that changes the velocity. You were comparing A&B to C&D. Gravitational acceleration requires more ropes (or effectively more ropes because of the fanning out).

A&B, if that's all there was, would be in equilibrium and so would C&D.

Of course, a two object universe is part of a thought experiment. There are ropes from every atom in existence connected to A&B or C&D. And there is a difference between objects falling in the earth's atmosphere and identically weighted objects being attracted to each other out "in" space, isn't there? But objects also reel in ropes and get closer together, or reel out and get farther apart.

Weight is NOT a force!

Rational Science destroys General Relativity's three irreconcilable mechanisms for gravity (wave, field and gravity well). First, we need to discuss the importance of a physical interpretation because math is useless to explain physical phenomena since it is purely descriptive. Besides, relativity is founded upon that which is not physical.

The mathmagician changed the meaning of physical interpretation when they had no real explanations for gravity and began to treat space as some kind of material that can be warped into the gravity well of GR.

Newton's force law can not be illustrated with a particle or a wave, gravitational waves can not "travel" at the speed of light and you can't curve space (because it is nothing). Yet, the inverse square factor of Newton's law of gravity says that as objects move closer then the attraction between two objects increases instantaneously and relativity says gravity travels at the speed of light.

A well is a photograph (static) whereas a wave is a movie (dynamic), but the Phiz Whizard will invoke whichever one supports the point he is making at that particular moment, and therefore has a moving target. So we can use the relativist's illustrations in order to explain the physical interpretation of General Relativity's gravity exposing the irrationality of it (fabric weighed down by an object). On the other hand, the rope architecture illustrates instantaneous action at a distance perfectly.

Of course their words are not scientifically defined, and one can never understand what it is that the Phiz Whizard is saying when he casts his spells. In physics we only use adjectives to qualify nouns (objects) so "finite speed of light" and "infinite speed of gravity" is neither grammatically correct nor rational.

This is why in science we use illustrations and define our Key Terms. Furthermore, we make an assumption and if the explanation contradicts the assumption we discard the hypothesis.

In RH the smallest unit of matter is the hydrogen atom, so we use two H-atoms connected by the rope. The rope is taut and tension (Mm/d^2) is constant because of the inverse relationship between frequency and wavelength ($c=f*Lambda$). It is the effective number of ropes which generates gravitational attraction. The tension is bi-directional and constant but there is no motion!

Tension implies that the two atoms are pulling on each other but the inverse relationship between frequency and wavelength insures a constant tension between them, not net force. Force is a one way mechanism and requires a movie to visualize ($F=ma$). Tension is a photograph.

Newton's equation: $F = G (m1 * m2)/d^2$

So what is distance squared mean? In the illustration of two objects A&B it means that every atom in A is connected to every atom in B. The ropes remain taut between the atoms.

Any atom of A that is between another atom in A and an atom in B is effectively neutralized. Think of two persons holding a rope at the opposite ends (representing to H-atoms and a single connecting rope). You hold the rope directly between them and they feel nothing, but pull the rope towards you (move the rope at an angle) and they feel you pull in your direction.

With many atoms there is the effective pull of many ropes. Tension between any two atoms is constant, but many ropes with different angles results in gravitational attraction (acceleration).

By changing distance between objects the atoms pull in different directions. Atoms along a single axis are in equilibrium so effectively only the two end atoms are pulling on each other. When the distance changes between two objects so does the effective pull.

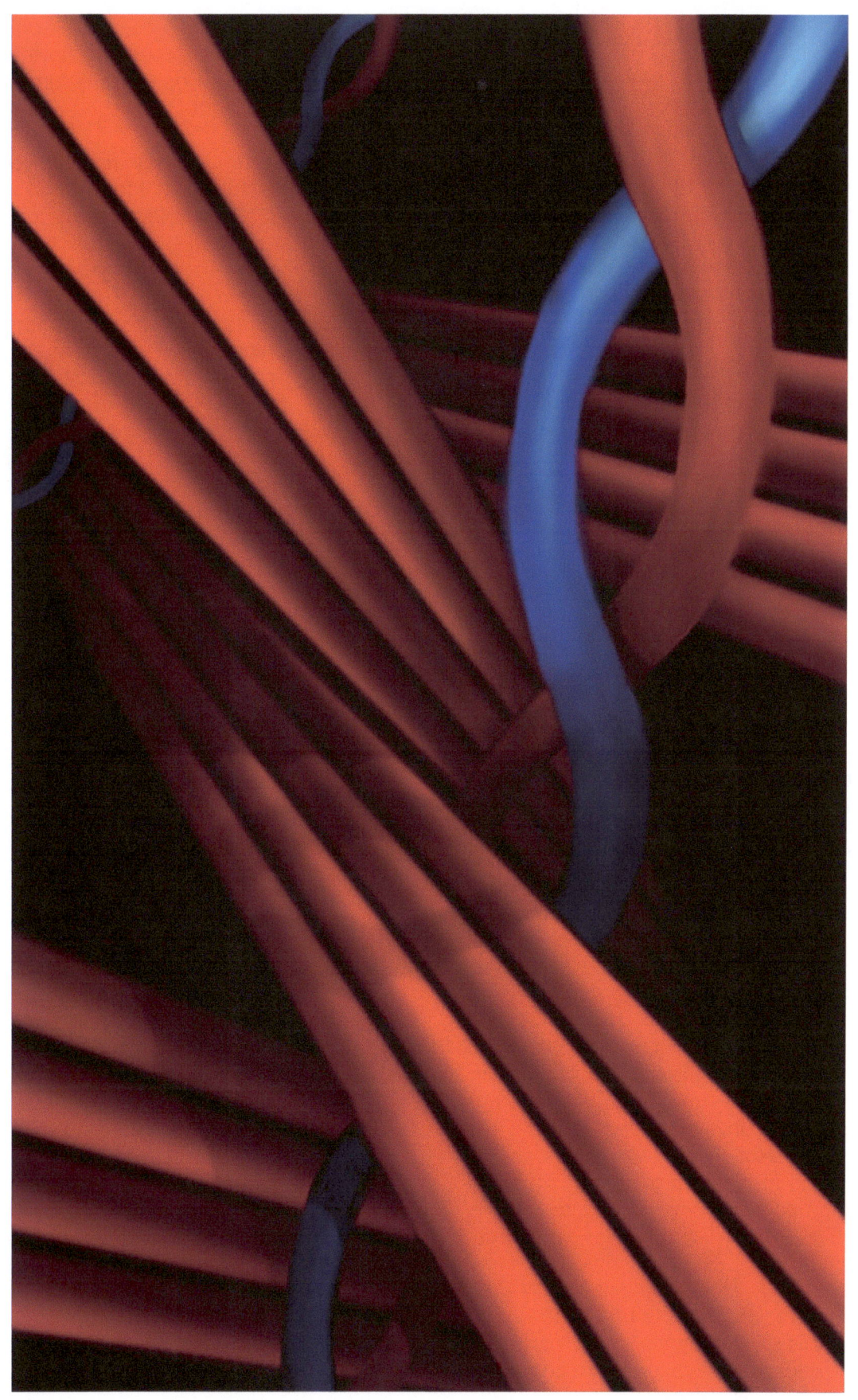

Light, Gravity and Magnetic Moment

Rational Scientist David Robison said, "Gravity falls off exponentially with increased distance. Light falls off exponentially with increased distance. Of course! Both light and gravity operate on EFFECTIVE ropes." This sparked an interesting discussion related to the Rope Hypothesis and how it is used in a theory of threads explaining the various phenomena.

Discussions of magnetic fields often refer to them as though they behave like light. "Does magnetic field strength follow an inverse square law?"

According to the Naked Scientist website, "Yes--the decay of $1/r^2$ comes from it basically being light-like electromagnetic radiation."

They are talking about "magnetic field strength around current carrying wires." A natural magnet's strength decays at $1/r^3$ instead of $1/r^2$. A hysteresis coil at a distance produces a magnetic field strength which decays at $1/r^3$, if I recall correctly from my TV repair days.

The difference is what they are calling monopoles and dipoles. One will often hear something like this from WIKI: "the magnetic field due to a dipole is inverse cubic, but the magnetic force from a monopole is inverse fourth order."

Magnetic moments, dipoles, monopoles; what does all this mean? Light and gravity conform to the hairless ape's "inverse square law" but magnetism to an "inverse cube law." Both are results of the Rope architecture, and since electricity and magnetism often are two sides of the same coin, we have to understand the relationship.

Whereas gravity "falls off exponentially with increased distance" because of "effective pull" and light "falls off exponentially with increased distance" because of changes in frequency and amplitude, magnetism falls off as the square or the cube depending on the source (current/monopole or magnet/dipole) and shape of wire/magnet.

Provided there is nothing between the observer/detector and the light source to change the frequency, the ropes still superimpose with distance (as with gravity) and the "effective light" decreases.

Current along a wire (read, enmeshed E-shells spinning in situ) along with the accompanying M-threads result in the field strengths detected. Speed of rotation results in a wider path of swinging threads. More magnetic molecules in a natural magnet, or more enmeshed E-shells (greater current) result in greater M-thread density. So, at near proximity to a source the numbers of threads per area are greater than distally.

The magnetic field strength detection and equations approximate the dipole fields falling off as the cube of the distance. Close in to a magnet, the proximal end can be detected/calculated and the distal end treated as if a monopole. In this case the field falls off as $1/r^4$ (quadractically).

Really close to one end of the magnet the fields effectively behave linearly.

Calculated or detected strength depends on location and shape of the source.

Reading the wiki article on magnetic moment and taking in their illustrations we see that an equation is used to calculate a magnetic moment. Magnetic moment is a vector quantity (magnitude and direction) derived as a ratio of torque to magnetic field strength.

From wikipedia, on magnetic moment: "The sources of magnetic moments in materials can be represented by poles in analogy to electrostatics. Consider a bar magnet which has magnetic poles of equal magnitude but opposite polarity. Each pole is the source of magnetic force which weakens with distance. Since magnetic poles always come in pairs, their forces partially cancel each other because while one pole pulls, the other repels. This cancellation is greatest when the poles are close to each other i.e. when the bar magnet is short. The magnetic force produced by a bar magnet, at a given point in space, therefore depends on two factors: the strength p of its poles (magnetic pole strength), and the vector l separating them. The moment is related to the fictitious poles as U=pl."

They use this as an analogy, but are closer to reality than they realize. The confusion for these guys is that they believe in spinning particles and electron orbits. They do not understand the relationship between E and M threads, although they observe (or calculate) the effects.

The relationship of magnetism to electricity is the serpentine drill bit's (enmeshed E shells) along with the accompanying jump rope like M-threads. For discussion about this with the luxury of detail, please go to the Rational Scientific Method or Rope Hypothesis facebook group.

Opposite charges are oppositely spinning E-shells along with the sweeping M-threads. Comparing the bar magnet to the electromagnetic loop in Wiki's image we see the same field strength and shape but at a different orientation.

Without an underlying physical mechanism like one the Rope Hypothesis provides, they are destined to observe, describe, measure and calculate forever without really understanding what's happening.

As long as they continually invent particles and orbitals they will forever be confused and doomed to stating things like this:

"The preferred classical explanation of a magnetic moment has changed over time. Before the 1930s, textbooks explained the moment using hypothetical magnetic point charges. Since then most have defined it in terms of Ampèrian currents. In magnetic materials the cause of the magnetic moment is the spin and orbital angular momentum states of the electrons, and whether atoms in one region are aligned with atoms in another." Of course, in order to "explain" these "point charges" and "states of electrons" they have to chant the mathemagical incantations of Quantum Magic.

Yes, they are very good at describing what they detect with their devices since the devices are calibrated and interpreted using their mathematical formulas. As the mountain said in "Me and My Arrow," "You see what you calculate to see and hear what you calculate to hear."

When they say this: "The net magnetic moment of any system is a vector sum of contributions from one or both types of sources. For example, the magnetic moment of an atom of hydrogen-1 (the lightest hydrogen isotope consisting of a proton and an electron) is a vector sum of the following contributions:

- the intrinsic moment of the electron,
- the orbital motion of the electron around the proton,
- the intrinsic moment of the proton."

What they are really describing is this: The effect of magnetic threads on our detectors is a calculated sum of thread density and direction of magnetic threads sweeping around an axis.

When you read this: "Similarly, the magnetic moment of a bar magnet is the sum of the contributing magnetic moments which include the intrinsic and orbital magnetic moments of the unpaired electrons of the magnet's material and the nuclear magnetic moments."
It means this: The atoms and molecules of a magnet align themselves increasing thread density sweeping in two opposite directions from a division we term north and South Pole.

Instead of this: "Viewing a magnetic dipole as a rotating charged particle brings out the close connection between magnetic moment and angular momentum. Both the magnetic moment and the angular momentum increase with the rate of rotation. The ratio of the two is called the gyromagnetic ratio and is simply the half of the charge-to-mass ratio."

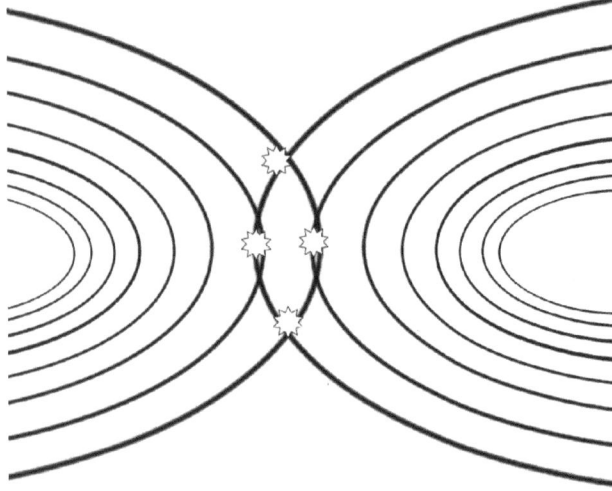

We can say this: A spinning atom's magnetic threads extend out from the atom's E-shell and sweep out around its proton's axis with more threads participating the greater the number of atoms involved. We can propose a ratio of M-thread density to speed of sweeping M-threads, and call it the gyromagnetic ratio.

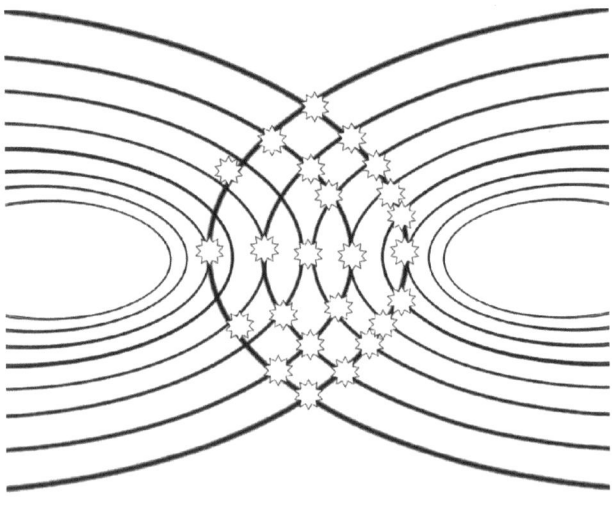

Conclusion: Whereas gravity and light "strength" are directly related to effective numbers of ropes, magnetic field strength is directly related to effective M-thread density and the speed of their sweeping action.

The shape of the magnetic source determines how many effective threads. Ropes superimpose over distance and M-threads combine or effectively cancel each other out depending on the shape of the source.

Gravitation and Electrostatics

Gravitational attraction between two electrons is 8.22×10^{-37} of the electrostatic repulsion at the same distance of separation.

Here's what they use to calculate the ratio between the two:

$k = 8.99 \times 10^9 Nm^2/C^2$,
$e = 1.60 \times 10^{-19}$ C

$G = 6.67 \times 10^{-11} Nm^2/kg^2$,
$m_e = 9.11 \times 10^{-31}$ kg

Both gravity and electrostatic "force" are stronger or weaker based on "the inverse of the distance squared." Even the two have similar formulas:

$F_g = G.M_1.M_2/D^2$

$F_e = K.Q_1.Q_2/D^2$

Similarities

"From looking at the two force equations, you can see the similarities and how gravitational force can be considered parallel to the force between two charges.

"Besides being proportional to the inverse of the square of the separation, both forces extend to infinity. They also both travel at the speed of light.

Differences

"One major difference is in the strength of the forces. However, gravitation usually is concerned with large masses, while any large collection of charges will quickly neutralize.

"Another difference between the two forces is the fact that gravitation only attracts, while electrical forces attract when the electrical charges are opposite and repel if the charges are similar. Thus, gravitation is considered a monopole force, while electrostatics is a dipole force." - From the School of Champions

Similarities are also used as an exercise in physics courses.

If Maxwell's equation ($c = \lambda f$) is the equation of a rope for light, then perhaps these equations are the equations of a rope for gravitation and electrostatics.

A single underlying mechanism for both phenomena makes it clear why the similarities and differences, are lost on folks like Franklin Hu who believe gravity IS "a straight forward application of the well known electrostatic force."

AND here we find this hilarious example of experimental logic from circlon-theory dot com:

"If we simply accept the most basic interpretation of this experiment, then we can easily determine what really happened in that apple orchard over three hundred years ago. After breaking loose from its stem, the apple remained motionless (except for a slight upward acceleration caused by air resistance) while the earth, the air, the tree, and Isaac Newton accelerated upward at the rate of 9.83m/s2, until Newton's head struck the nearly motionless apple as it floated upward on the rising column of air."

Instead of writing 2 equations, simply draw a two atom universe based on the rope model and discover the inverse relationship for electrostatic "force." Draw a 2 object universe with multiple atoms at various distances and see the inverse relationship for gravitation.

What does this describe in regards to the rope model? In other words, what about the rope architecture insures that electrostatic repulsion is greater than gravitational attraction at this micro level, but gravitational attraction exceeds electrostatic repulsion at macro levels of MonkEs and men, stars and Galaxies?

The simple answer is that at the macro level there is little or no matter between any two objects for enmeshed E-shells to produce current or, individual M-threads to interact, but there are still ropes connecting all atoms. At the micro world of, say two adjacent H-atoms, there is only the static tension between the two by virtue of the connecting rope, but E-shells can easily enmesh and magnetic threads can interact.

There are many other interrelations at the level of the fundamental unit of matter.

The theoretical physicist reifies phenomena into particles. The Rope Hypothesis provides an underlying physical mechanism by which Thread Theory can explain all phenomena. Rope Hypothesis provides a model by which we may unite not only gravity and electrostatics, but magnetism and light.

Let's take a look at how these various phenomena are related.

Particle physicists tell us that most of the mass of an atom is attributed to the nucleus. With the rope model, the E-thread extends from each H-atom to the center of all other H-atoms. There will be as many M-threads as there are E-threads contributing to the E-shell. Since electrons are actually NOT discrete, but layers of the E-shell, and M-threads sweep out from the atom, the

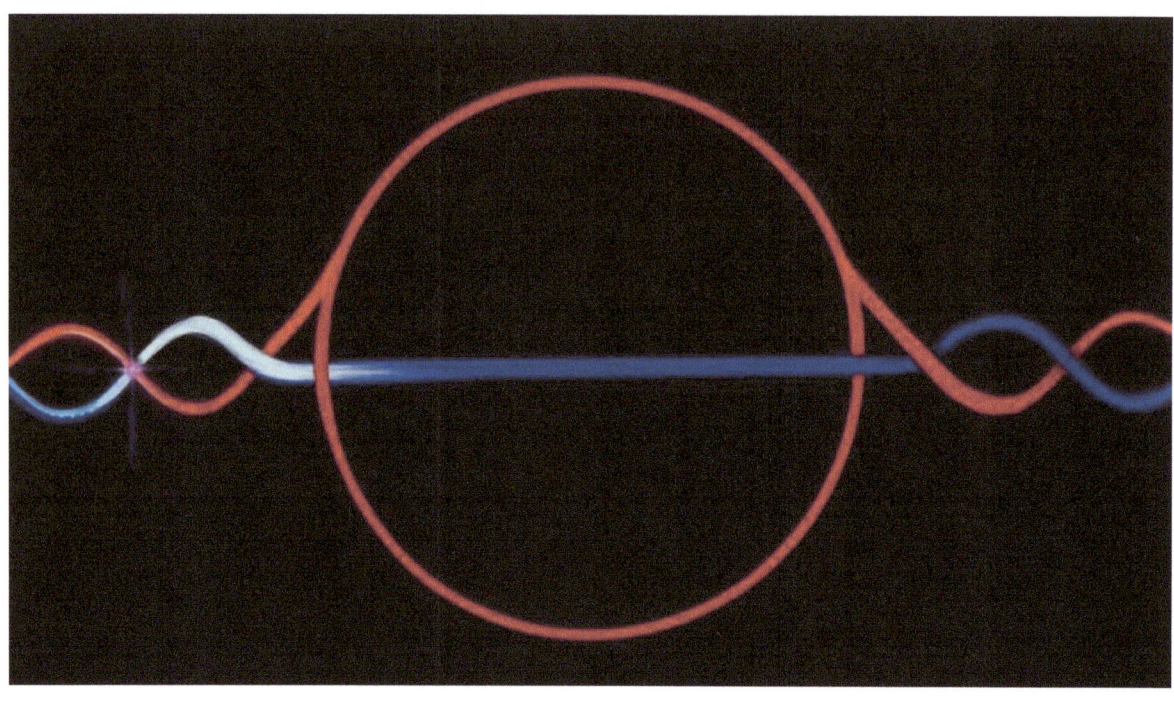

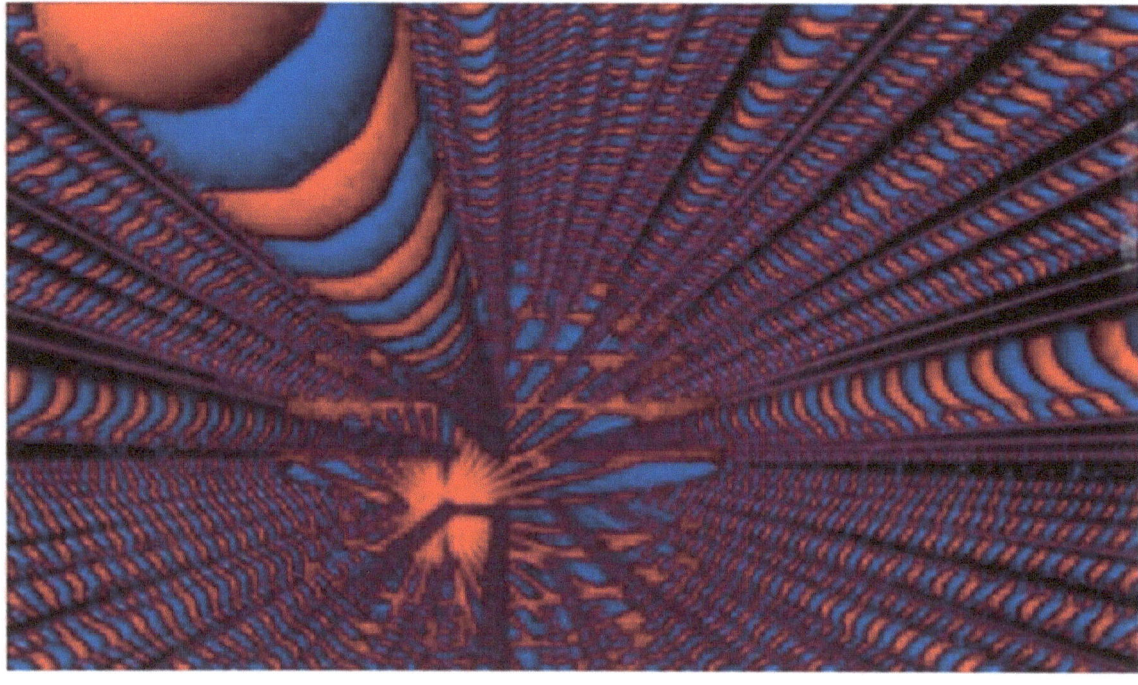

diameter of the H-atom is about 145,000 times larger than the nucleus, but the particle theorist attributes more mass to the nucleus.

Listen to this from Wikipedia:

"The nucleus is the center of an atom. It is made up of nucleons (protons and neutrons) and is surrounded by the electron cloud."

No wonder they attribute most of the mass to the nucleus, they think the E-shell is a cloud!

If thread is reeled out in one place, slack in thread must be taken up somewhere else. There is a finite amount of thread after all. This can explain inertia and acceleration. What if the slack is being taken up in the sweeping M-threads? In that case, the E-shells would constitute a greater amount of thread than the proton. There is an interrelation of the atoms in any given object. In other words, there is a ratio of H-atoms to other atoms. Perhaps this ratio is maintained by varying pump, spin layers, or by the ratio of pump, spin, and slack between various atoms.

Layers: thread or threads donated from rope or ropes.

Usually, when we refer to mass we mean the number of atoms comprising all matter, (matter being the aggregate of all atoms). Mass for an object is the amount of matter comprising it. Mass in terms of Newton's equation is a result of the pull of every atom in the universe via the interconnecting ropes. Here we are engaging in mereology, as well as using our term "mass" ambiguously. This is a result of trying to explain an object, especially the fundamental object which is comprised of thread, rope, and atoms as though they were separate entities. Sub-atomic discussions necessarily result in this if there are no discrete particles. We don't explain objects – we point to them and name them, we explain phenomena. However, we are trying to understand what the Particle Phiz Whiz is describing when they invoke sub-atomic particles. They describe phenomena and explain objects. Let's be careful not to do the same. We can use the rope model to explain all.

It is said that the electrostatic and gravitational forces are not confined to macro objects but also between the electrons and nucleus of a single atom.

The electric force is related by Coulomb's Law as discussed in detail in the Rational Science Vol. V, and Rope Hypothesis and Thread Theory books. We leaned that an ion is an atom or molecule with a net electric charge due to the loss or gain of one or more electrons.

AND we understand charge is not only related to numbers of particles but is inherent in sub-atomic particles, as discussed in the chapter Elementary Charge.

With the rope model, charge and elementary charge can be related to direction of spin of atoms, and direction of spin of layers of atoms.

Particle physics is in a quandary:

The particle phiz whiz measures the magnetic moment (magnetic field) of an electron and "knowing" its "charge" figures the rate of spin. But this poses a problem. In order for the electron to have the magnetic moment that it does it must be spinning faster (actually called tangential velocity) than light. Since faster than light (ftl) is not possible they conclude that the electron does not spin this fast. In fact, they say, angular momentum alone accounts for the magnetic moment. So, unlike other spinning bodies, sub atomic particles don't actually spin to produce these magnetic fields.

When particle physicists say, "angular momentum alone accounts for the magnetic moment" they are talking about its moment of inertia and its angular velocity. This could actually be one layer of an E-shell, or an E-shell, and would solve their quandary.

We assume that voltage is a result of the number of enmeshed E-shells, and current is a result of the rate of spin of these E-shells.

Magnetic strength is a direct result of M-thread density due to alignment of molecules. M-thread density is indirectly related to the rate of spin, as current is related to the rate of spin. How fast an E-shell, or layers of an E-shell are spinning may determine how far out the M-threads sweep.

We see that magnetic strength is related to the alignment of molecules in a magnet, but an electromagnet has different properties because of the shape of the sweeping M-threads depending on whether "dipole or monopole" structure of the magnet. This is related to shape and nature of the "magnetic field," which is determined by the shape of the magnet (natural or coil or straight wire).

We consider that G (tension) averages out universally by virtue of a constant number of H-atoms, and c (torsion) is also constant because of the inverse relationship between frequency and wavelength which also depends on the constant number of H-atoms. G is a universal average, but gravity locally is a result of the effective pull of ropes.

Coulomb's Law states: "The magnitude of the electric force that a particle exerts on another particle is directly proportional to the product of their charges and inversely proportional to the square of the distance between them. The direction of the force is along the line joining the particles."

As we have proposed, current flowing past a certain point is really a count of how often an E-shell is spinning on its axis at a given point and in a given amount of time.

An individual "particle" would then be a single E-shell or layer of an E-shell. The magnitude of electric force, one layer on the other layer, is directly proportional to their product. The "square of the distance" holds because the E-shell is comprised of a thread, or threads donated from a rope, or ropes whose relationship between frequency and wavelength is inversely proportional. "The direction of force along the line of particles" is by virtue of the E-shells being comprised of a single thread.

The architecture of the rope model insures that no matter what you are talking about something is going to relate either directly or indirectly, and/or either proportionally or inversely proportional to something else! Hence we note gravity and light, electricity and magnetism have the aforementioned relationships.

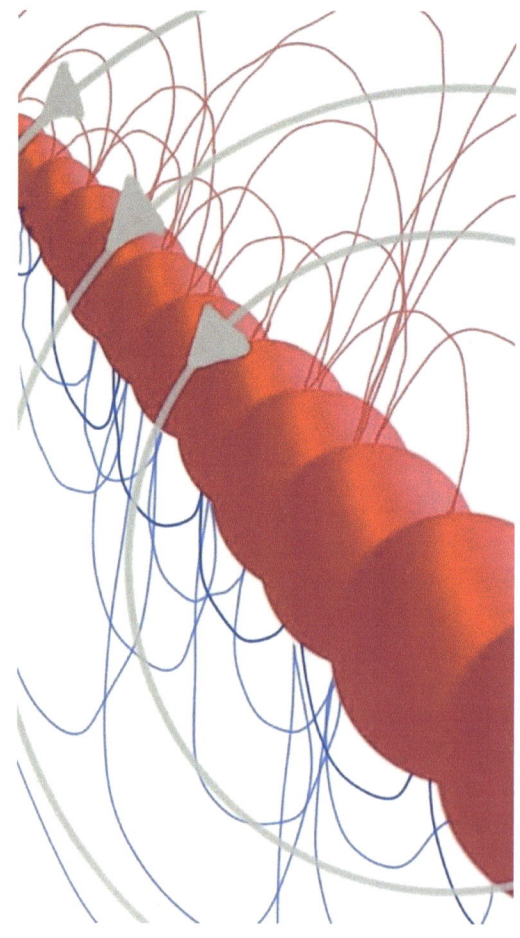

Photoelectric Effect and Thread Theory

In this article we will examine, however briefly, the photoelectric effect as it relates to the Rope Hypothesis.

The explanatory power of the Rope is unsurpassed in its ability to model reality. There is not a single phenomenon which can not be explained using the rope architecture. Still, this is not why we reject particle physics... not because we think that the alternative is better. It is because the particle/wave paradox is impossible.

We don't reject the mainstream's "explanation" for the photoelectric effect because we think there is an alternative. We reject the "explanation" for the photoelectric effect by particle physics because it is NOT an explanation at all. Please review the Rational Scientific method.

One may claim that the photoelectric effect proves or confirms photons all they want, but photons can not possibly exist! This is covered in great deal in the Rational Science series of books found on Amazon.

"The photoelectric effect is the observation that many metals emit electrons when light shines upon them. Electrons emitted in this manner may be called *photoelectrons*.

"According to classical electromagnetic theory this effect can be attributed to the transfer of energy from the light to an electron in the metal. From this perspective, an alteration in either the amplitude or wavelength of light would induce changes in the rate of emission of electrons from the metal. Furthermore, according to this theory, a sufficiently dim light would be expected to show a lag time between the initial shining of its light and the subsequent emission of an electron. However, the experimental results did not correlate with either of the two predictions made by this theory.

"Instead, as it turns out, electrons are only dislodged by the photoelectric effect if light reaches or exceeds a threshold frequency below which no electrons can be emitted from the metal regardless of the amplitude and temporal length of exposure of light." - WIKI

One would reasonably think that if theorists are going to rely on experimental results to confirm their theories then they would have thrown out the theory when their experiments did not confirm it! Instead, they just added particle theory to wave theory and light became a wavicle. It behaves like a wave sometimes and behaves like a particle other times. In fact, they say, it leaves as a particle, travels as a wave and arrives as a particle.

The experiments all claim that electrons are ejected from metal surfaces when bombarded by light. But when one takes a look at the experiments, they only show an effect that is claimed to be caused by the ejection of electrons. How did they arrive at the conclusion that electrons are being ejected? Math!

What the experiments revealed:

1. Energy of electrons increases with light frequency.
2. Current remains constant as light frequency increases.
3. Current increases with light amplitude.
4. Energy of photoelectrons remains constant as light amplitude increases.

So, they measure current, calculate energy and then, based on the math, they extrapolate that electrons are being ejected.

The experiments haven't really changed much as you can determine from this UCLA physics experiment:

http://demoweb.physics.ucla.edu/content/experiment-6-photoelectric-effect

They use a photodiode and an amplifier, batteries, a voltmeter, a light source and a filter to vary the light intensity. Shining a light on a metal surface and measuring the voltages at varying frequencies and intensities, to them, verifies the theory that particles are responsible; same results, just a new theory. There are basically two components here: voltage level measured, and various results due to light frequency and intensity.

There are other simpler demonstrations you can view on YouTube showing the "photoelectric effect" using static charge to charge tinsel and then ultraviolet light to discharge the tinsel.

https://www.youtube.com/watch?v=muxRZ1irsrk

Are these experiments demonstrating that electrons are being "ejected?" Only, if there are light particles called photons and particles called electrons, charge, etc. What does this really show? It shows a simple case of confirmation bias.

As for amplitude, frequency thresholds and work function, that is another matter, and only secondary to the issue of electrons being ejected from metal surfaces. The mechanism is related to the amount of energy (hence frequency) required per given element which increases as the atomic number increases. Frequency, amplitude and Planck's constant are covered with detail in Rational Science Vol. V.

As we have already learned from the Rational Scientific Method, experiments are extra-scientific, and anyways, without a viable hypothesis they have no theory. Since waves and particles fail at the hypothesis stage they have no theory to begin with!

Particles have no way of explaining why higher frequencies result in higher energy. And the higher amplitude light beam does not represent more photons as is suggested, because there are no discrete photon particles.

Energy is the magical word being used here. We are told that higher frequency somehow imparts higher energy to kick out the electrons faster and faster. However, we learned early on in grade school that energy is the ability to do work. Now, these deluded souls wish for us to believe that "ability" is transferred from a zero dimensional particle (photon) to discrete electron balls in the metal plate.

Zero dimensional (photon) particles can not possibly exist and "ability" can not be transferred, so the particle physicist needs to go back to the drawing board and he or she also needs to abandon his or her extra-scientific experiments.

But let's briefly look at what is possibly happening using the rope model of light; the same model that explains electricity, magnetism and gravity.

Since all atoms in the light source are connected to all the atoms in the air and in the metal plate, when the light source is turned on it induces the atoms to pump at a rate which results in friction (charge) along the threads of the electron shell. The ropes twist (torsion signal) and also reel in and out between atoms. This is what is known as light.

The number of links per unit length is what is referred to as frequency. The height of individual links is amplitude.

The ropes between atoms whose electron shells are being induced to pump faster have a greater number of links per unit length. Higher amplitude is a result of the superimposition of ropes converging on atoms.

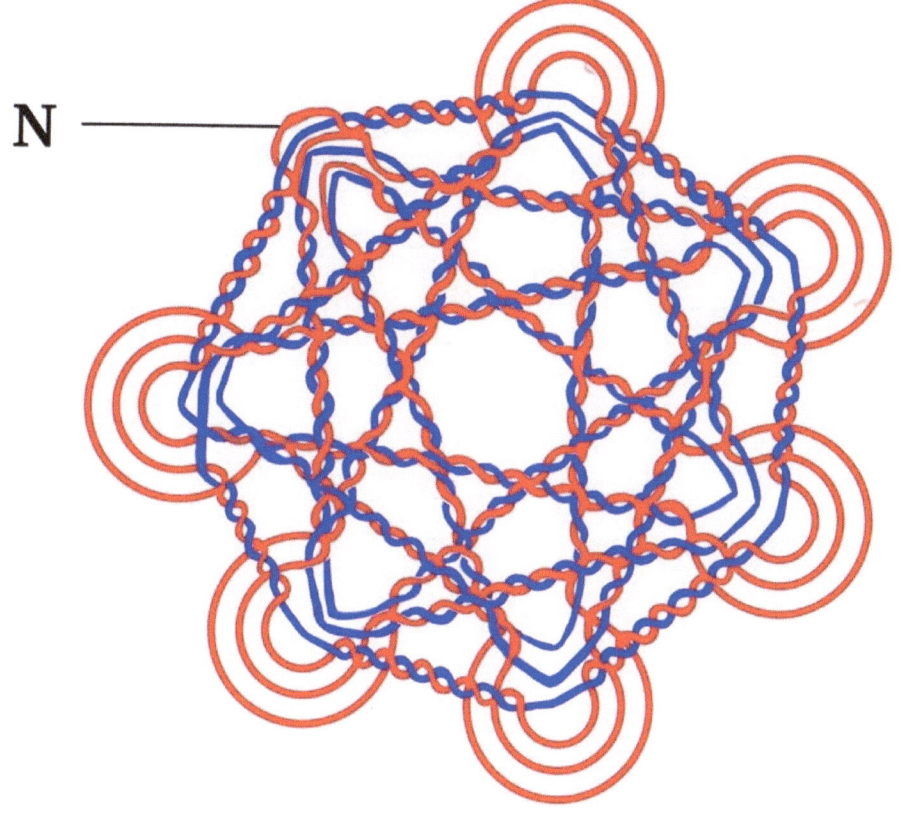

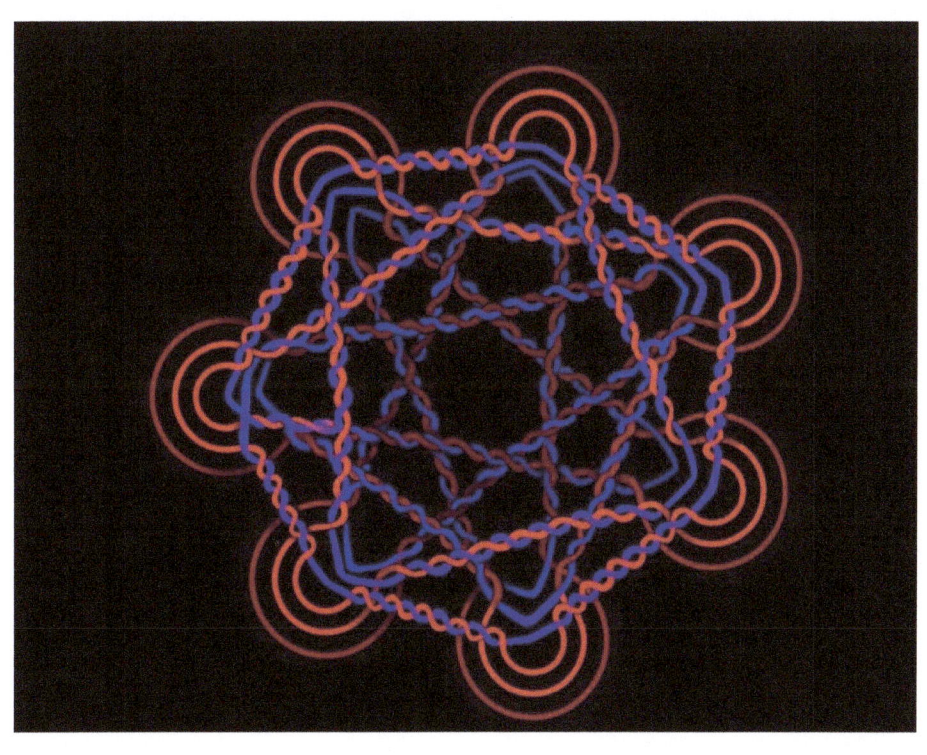

Increased Amplitude: Same number of links with higher height to each individual link. Increased Frequency: faster pumping atom but same amplitude with greater number of links per unit length.

Light shining on the polished metal surface induces electron serpentines to spin. Electron serpentines are enmeshed E-shells of adjacent atoms which turn together like a drill bit, and this electrical current is measured or "felt" in other conductors.

So there you have it, the photoelectric effect without any irrational and impossible, massless zero-dimensional photon particles.

How is Sound Different than Light?

There is something about different wavelengths that determines how light and sound interacts with matter. X-rays interact differently with matter than light waves, etc. In view of the Rope Hypothesis, it might be better to say that matter interacts with matter manifesting in different ways in "portions of the electromagnetic (EM) spectrum."

All atoms are connected to all other atoms by a two strand EM rope. Regardless of which slice of the EM spectrum we are considering, all wavelengths are a result of the number of links per unit length of each rope connecting any two H-atoms. Different materials have atoms interconnected with ropes of different frequencies and wavelengths.

Martin says, "We can hear around a corner, but we cannot see around a corner. My guess: Sound is affecting connected matter whereas light is already in connection with everything."

I can shine a laser at your window and listen in on your conversation, because sound waves interact with light waves, and I can translate those light waves into sound waves. Why is this?

ALL light is bi-directional and rectilinear in propagation as it travels along straight ropes. Let's be clear that we are talking about light as in "all of the EM spectrum," but visible light is only a small range of frequencies and wavelengths. Since atoms are all interconnected, atoms from a light source "around the corner" are also connected to the brick building and to our eyes. We do not see the light because the brick and mortar has shifted the wavelengths out of the visible spectrum. X-rays, another form of light, pass right through the brick and mortar and can be detected by X-ray sensitive equipment.

Listen to this guy's explanation from a blogspot dot com blog entitled sound and light reflection: "Sound can also be carried by other media, such as water (e.g. sonar). Light, on the other hand, is an electromagnetic wave, which means that it is a combination of electric and magnetic fields that travel together. Therefore, it does not need air, water, or any other medium to carry it - in essence, it carries its medium with it."

This is the kind of convoluted nonsense we hear from quackademia. How does a wave carry its own medium? A wave is what something is doing. A field is where something is located.

Phiz Whizes, like him, think that sound IS a longitudinal wave and light IS a transverse wave, and that light is made of two transverse waves. He will tell you that a transverse (light) wave travels a million times faster than a longitudinal (sound) wave. But seismologists will tell you the longitudinal waves travel almost twice as fast as transverse waves during an earthquake. Clearly we have to consider the medium involved in either sound or light. What does that tell us, that is, compression waves traveling faster than shear waves? The former compresses earth in the direction of travel whereas the latter shakes the ground back and forth perpendicular to the direction of travel.

As we look at the left side of the EM spectral chart, we see lower frequencies (fewer number of links per unit length). As we move to the right, higher frequencies (higher frequencies smaller wavelengths) sound, radio waves, TV, microwave, visible light, X-rays and cosmic waves.

In communications, lower frequencies in the 1Hz (sub sonic), are used to send signals through the earth, whereas higher frequencies, in the radio and microwave range, are used to send signals through the air.

We are told that the speed of sound is independent of frequency. Speed of sound depends on density, but also on compressibility and something called the "shear modulus." Since sound travels longitudinally, solid materials with higher compressibility are a better medium for sound. Clearly, matter matters.

Remember I said this? "There is something about different wavelengths that determines how light and sound interacts with matter."

What I really meant was this: "There is something about matter that determines wavelength for light and sound."

Sound, visible light and X-rays are all the result of torsion along ropes between atoms and whose frequencies and wavelengths vary.

Light & Sound, How are they alike?

Mixing primary colors gives us secondary colors, but no mixing of secondary colors can give us primary colors. Why is that?

It is the difference between fundamental and composite "waveforms." A composite waveform is merely multiple ropes.

The color green has a wavelength of ABOUT 510nm. What one perceives as green is actually a range of wavelengths (Green 577 - 492). BUT, let's select 510nm as the fundamental color "green."

Add 475nm (blue) and we have a composite color. Let's call it aqua, or blue green.

Add 510nm to 510nm and we have more intense green (more ropes with the same number of links per unit length), but it's still green.

Sound is similar to light in respect to fundamental and composite waveforms, but there is a difference between light and sound in how we perceive it, and not just because our ears are different than our eyes.

Humans really only have one sense, and that is touch. Surface to surface contact between molecules of air and the sensory apparatus "ear" stimulate electro-chemical reactions in the auditory cortex resulting in what is known as hearing.

Surface to surface contact between molecules of various substances and compounds with two patches of sensory cells in the nose relay information by olfactory nerves to an area of brain that converts those impulses into something called smell. Surface to surface contact between food and the tongue produces a brain response called taste, and light "touched" rods and cones on the retina in the back of the eye result in something called vision which is experienced in the visual cortex of the brain.

The sense called touch is surface to surface contact between objects and receptors on the skin, muscles, bones, joints, and organs. There are chemical, temperature, mechanical, position, and pain receptors that relay information to the parietal lobe and the cerebral cortex. Taste combines tactile, auditory, and chemical cues.

All these senses are a result of surface to surface contact between objects, so in a way, we really only have one sense; the sense of touch. We shouldn't be surprised. We understand that ALL phenomena are the result of surface to surface contact between two or more objects. There are only two kinds of "touch": push and pull.

Light is torsion along ropes simultaneously in both directions. Sound is a form of light, and so it too is torsion along ropes simultaneously in both directions. BUT, isn't light a form of touch known as pull. What about sound? Isn't this a form of touch known as push? I clap my hands, and the air is pushed forward towards your ears. The air molecules are directed into your inner ear and push against the tympanic membrane.

The problem lies in using the word TOUCH. What does this really mean? Certainly the particle phiz can not tell you that there is no distance between the air molecules and the tympanic membrane. If this was the case, then the air and membrane is one object, and if there is only one object then we can't have touch. BUT this is only the beginning of the quantum mechanic's troubles.

We are told about the "electrostatic repulsive force" between electrons, or graviton balls imparting negative momentum. In the first case, what prevents negative electrons in air atoms from attracting positive protons in the tympanic membrane atoms (the electrons are not always on the same side of the proton). In the second case, how do colliding particles pull on each other without an intermediary? Nonsense!

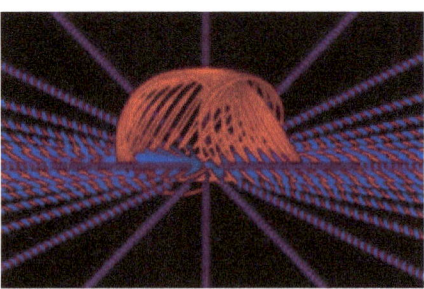

On the other hand, if all atoms are interconnected by EM ropes, we can explain both push and pull. Sight is a result of both push and pull, but hearing is, and the rest of the senses are, a result of push. In the case of sight, torsion is traveling along ropes between atoms in both directions, and with sound push happens directly between atoms as the electron shells come into contact with each other.

We often hear something like this, "Light does not require a medium to travel, but a longitudinal wave cannot propagate without a medium." We are told that sound travels longitudinally in gases but we find it can also travel transversely in solids. Actually nothing happens without a medium including light. So, what are they talking about?

- A longitudinal wave is a wave vibrating in the direction of propagation.
- A transverse wave is a wave vibrating at right angles to the direction of its propagation.

Of course, both visible light and sound is torsion along ropes, but in order for us to hear a sound there needs to be a perceived difference between the compression and decompression of the medium which is happening at right angles to the direction of propagation.

Light is a two way mechanism by virtue of interconnecting ropes between atoms. Sound is not returned directly back to the source because displacement of the medium is perpendicular to the direction of propagation, and, remember, this requires E-shells to interact with E-shells.

Remember I said this about the difference between sound and light?

"It's the difference between fundamental and

composite 'waveforms.' A composite waveform is merely multiple ropes."

Mother Nature equipped us with a specialized sensory apparatus with limited bandwidth. Since every atom is connected to every other atom, everywhere, we are receiving (and transmitting) torsion signals between every atom in our body and every star in our galaxy. In other words, it is the particular composite waveforms we are aware of that inform our experience. If we didn't have limited bandwidth sensory organs, our brains and bodies would be overwhelmed with billions of frequencies and wavelengths and amplitudes and unable to distinguish green from blue, let alone light from sound.

Fortunately, we have the unlimited ability to conceive of concepts and can explain all phenomena and even the underlying invisible physical mechanisms. AND we can do this consistently with a Theory of Threads.

Do you suspect that something is wrong with what you read about in the so-called physics journals and popularization magazines? Don't black holes sound a bit magical? Doesn't Big Bang come across as Creationist? Does the establishment's "explanation" that there are many copies of you, each in a different universe, make you wonder about the state of mind of the theorists? If so, perhaps it is time for you to consider an alternative.

Why God Doesn't Exist (WGDE) presents a fresh perspective on the nature of light, gravity, magnetism, the atom, and the workings of the Universe in general that is rational. It is rational because each theory is illustrated. You can understand the mechanism by just watching the videos that accompany the book. It is also mainstream theories, you will not be asked to believe in the movement of concepts (e.g., transfer energy, move "a" mass, carry "an" interaction). WGDE is for intelligent laymen who simply want to understand the causes and mechanisms that underlie physical phenomena.

To obtain a paperback, Paypal to bill@youstupidrelativist.com
USA/Canada: US$ 30   Europe: 30 Euros

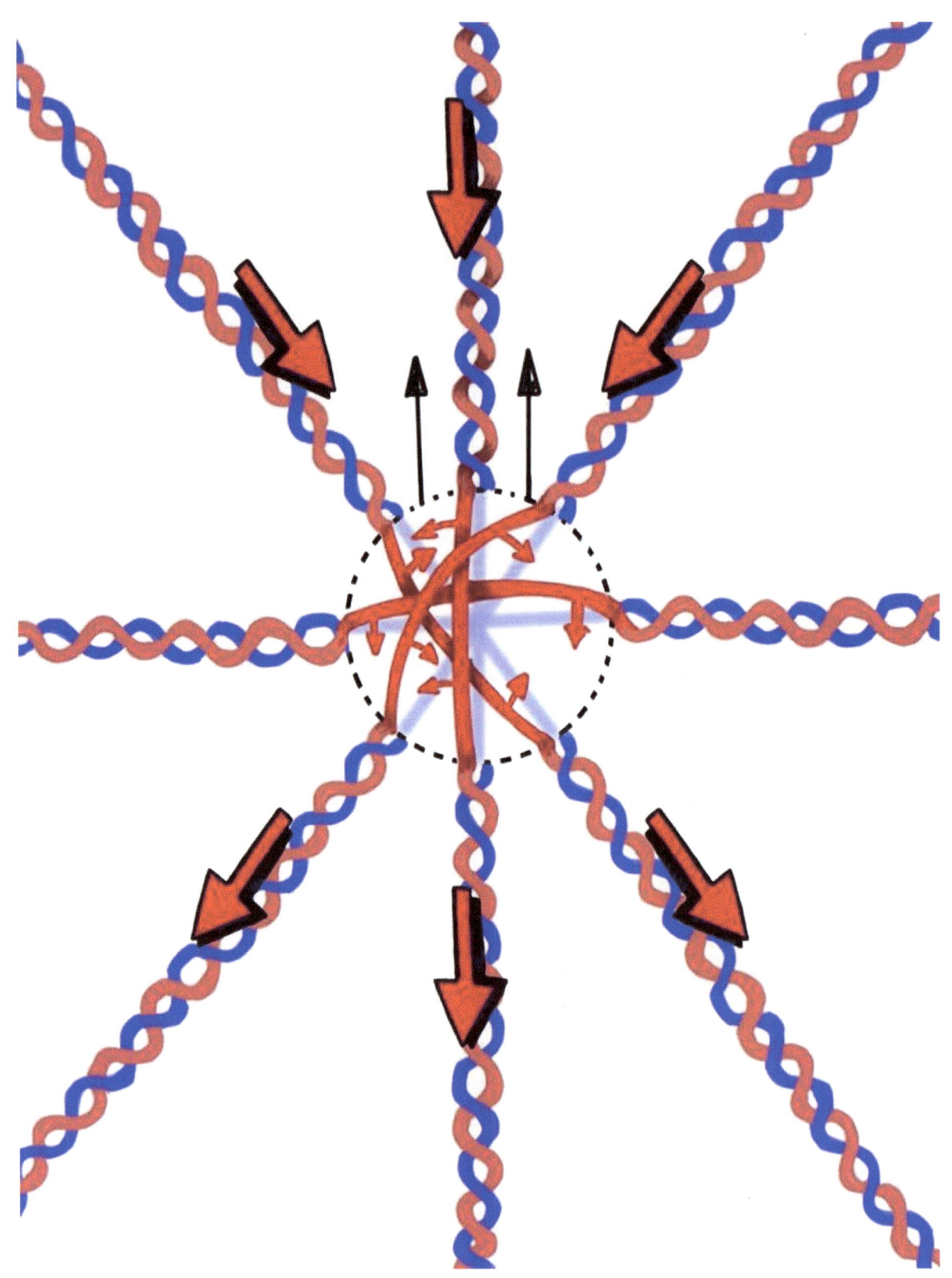

Rope Hypothesis
and
Thread Theory

By
Monk E. Mind

Refraction, Reflection and Diffraction

Let's review some basics before going into detail. What is refraction? How do we explain differences in velocities of light passing through various media?

Snell's Law doesn't explain why the refraction of light is different through different media. Only understanding the underlying physical mechanism will shed light on the subject.

Light is not bending; it is taking a different rectilinear path (a different angle). Light is a torsion signal along twined EM threads forming ropes interconnecting all atoms.

Let's shine a light beam through a piece of glass. Atoms in air and glass pump at different rates and so the interconnecting ropes have different numbers of links per unit length (frequency).

Our light beam induces atoms in the air to torque faster. Glass molecules reflect some of the torque signal and refract some of the torque signal.

But what does this mean? Atoms of the air interacting with atoms of the glass cause the glass atoms to pump faster. Since glass and air atoms pump at different rates and the corresponding frequency is different, the molecules of glass act as a sort of shock absorber re-transmitting the torque signal to other molecules of glass resulting in the different rectilinear paths known as reflection and refraction.

What is reflection? Is it like Newton's bouncing ball? Is it Huygens's wavefront bouncing back? Neither particles nor waves can explain how a laser light can be bounced off of mirrors on the moon and return to its origin a couple of seconds later here on Earth. Only if light travels rectilinearly along a path in both directions can this feat be accomplished.

What is diffraction? Is it light bending around corners, and waves spreading around obstacles? No, light is neither particle nor wave, light does not bend, or slow down or speed up.

For the real scoop on light see Rational Science Vol. I, Chapter Eleven, "Light, Particle Or Wave?"; Rational Science VOl. II, Chapter Fifteen, "The Nature of Light," Chapter Sixteen, "Light ...Does It Travel Rectilinearly or Curvilinearly?", Chapter Seventeen, "Distance To the Stars," Chapter Eighteen, Shapiro Effect and Chapter Nineteen, "Distance ...The Rubber Ruler"; Rational Science Vol. IV, Chapter Thirty Five, "Why Fattie Gets A Sunburn," Chapter Thirty Six, "Backscattering," Chapter Thirty Seven, "Neutron Bombardment, Beta Decay, and Radiation" and Chapter Thirty Eight, "Photoelectric Effect and Rope Hypothesis."

Now, sit in on a conversation with rational scientists as we discuss light phenomena.

David Robison says, "Still don't understand the directional preference of reflection, refraction and diffraction. Why wouldn't the immediate light from the Sun overwhelm the coronal atoms and cause them to pump according to it? Can a single atom relay multiple signals at once? As in magnetic threads pumping differently within an atom?"

"Directional preference" has to do with atoms relaying the signals to your eyes in a frequency and at an amplitude that your brain can perceive and your eyes can handle. You can not stare directly at the sun, because of the amplitude of the light signals (height of links). You can block out the sun and view the corona.

Atoms in the sun and atoms in the corona are connected to your eyes. So, there are multiple paths to your eyes. But this is a separate question from do atoms relay

The rope relays the signals, not individual threads. The pumping of the atom torques the rope. Can there be "ripples" or harmonic resonances from individual threads superimposed on the rope between two atoms? Why not?

Paul says, "Only atoms with multiple E-shells can do this. Poor old hydrogen can only do one thing at a time."

Is tension, vibration, or torsion "felt" along threads at all? Or, is this phenomena restricted to ropes? Just as my swinging through the jungle has little effect on Andromeda because of its tug-of-war with the Milky Way, do X-rays, Gamma-rays and Manta Ray's originating in Andromeda have little "noticeable" effect on me because of torsions originating in the Milky Way and solar system?

By "felt" and "noticeable" I do not mean to put human observation front and center. I'm just interested in the effects, consequences, of these phenomena.

As Paul points out, EM threads that comprise H-atoms are arriving from every other H-atom. All other atoms are comprised of H-atoms, so they have merged E threads, and multiple E-shells. What about contributing threads? How is torsion relayed?

What is refraction? Here is a typical definition:

"Refraction is the bending of a wave when it enters a medium where its speed is different. The refraction of light when it passes from a fast medium to a slow medium bends the light ray toward the normal to the boundary between the two media." - From Hyperphysics, hosted by Georgia State University.

Bending of a wave? The speed of light changes through different mediums?

Why doesn't refraction occur in space? What makes light return back to c after being slowed down through a medium? AND according to Snell's index of refraction: "The index of refraction for a substance is equal to the speed of light in free space divided by the speed of light in that substance."

If light slows down through a medium and speeds up through free space it is also speeding up and slowing down within all mediums as it encounters atoms and "empty space." Amazing! Refraction doesn't occur in free space because it either doesn't encounter any medium (few atoms in empty space). BUT WAIT! I thought space IS a medium.

Newton thought that light would speed up through denser medium (like sound does) and Huygens thought that light would slow down. Newton believed in particles and Huygens believed in waves. How do they measure the speed of light anyways? Turns out they don't. Mathematicians measure angles and then calculate velocity.

David says, "I don't know how light would manage to pass through that boiling bubbling brew of virtual particles known as free space without slowing down at all."

If it does, we'll never notice because it speeds up again in free space, and light is coming at earth from all direction simultaneously. Be kind of hard to trace a single wavicle, wave or particle from a single source. How convenient for the theorist!

Speaking of light speeding up and slowing down:

The laser ranger station near me reflects a laser beam off of a mirror on the moon and times the trip in order to measure the distance to the moon. We are told that light leaves as a particle, travels as a wave, and then arrives as particles where they are detected by a photon detector.

What? Depending on the amplitude of the waveform, a wave packet could take a path as much as 2x further or more. Therefore it would have to travel twice the distance in the same time or in other words, the wave packet would have to travel twice the speed of a single particle traveling a straight line path at the same distance. But photons travel at c. Light can not travel faster than c. Therefore, light can not travel as a wave packet, or the detector is not really detecting individual photons as claimed.

Also, if light does not travel a straight path, how does the light return to the Laser Ranging station over 2 seconds later when the earth has already moved many kilometers?

A typical answer is that the photon doesn't move faster than light, it runs in place while the wavefront moves at c (except, of course after it is refracted by the earth's atmosphere).

David: "Not to mention the poor photon has to navigate warping curving spacetime filled with cosmic potholes and walls and the like. He'll be rolling and bouncing all over the place."

Wladimir says, "Another pointer that space is indeed not a medium is that of impedance matching. In order for light not be reflected and actually go through a medium, the thickness of the body in direction of the path must be multiples of $\lambda/2$. In the words of my professor: Whenever light is reflected at a surface of a material depends entirely on the path length the light "sees" from the entry point. Meaning what happens at the interface is dependent on what is beyond that interface.

"This is one of the most counter intuitive concepts of electrical engineering and one physicist have a hard time to grasp. This result is derived through transmission theory (transmission of electrical power through a wave guide) and is routinely used in optics (if you ever wondered how anti-reflective coating works, this is it).

"Space does not need any impedance matching. It is not a wave guide! Electrical engineers refer to the propagation of an electromagnetic 'wave' through a medium as the 'electrical path length,' differentiating it from the propagation through 'space.' In short: Propagation through a wave guide is different than through 'free space.' In simple terms: The ray is bouncing inside the medium that constricts it and the resultant path will necessarily be longer than the direct path between entry and exit."

Interesting idea. I've never made the connection between impedance and reflection of visible light before. And of course, what happens at the interface is dependent on what is beyond that interface because all atoms in the interface are connected to both the light source and atoms beyond the interface.

BUT back to refraction. What is it?

Refraction is, simply put, the changing of angles of the connecting ropes between objects. Not really ropes changing angles, torsion signals being felt along different ropes (incident light) which have different angles then the so-called "normal path."

So why does it appear that light is speeding up and slowing down? Because calculations of speed are being made based on angles of incident light rather than actually measuring speed.

Anyone care to explain this with the luxury of detail? According to the YouTube video entitled Sixty Symbols by Professor Merrifield, the refractive index doesn't depend on the wavelength of light as one would expect. Since we "know" that atoms absorb light at particular frequencies, and a material has many atoms, there should be random encounters between light and atoms and so not always the same refractive index for the same material.

The classical answer, as professor Merrifield gives it in the video is that we have the superposition of waves caused by atoms responding to the EM fields of the light waves and producing their own EM fields as waves which somehow accumulate and are seen as one basically coherent wave pattern exiting the material which is now traveling less than c.

In other words, combining many waves propagating at c ends up producing a composite waveform which somehow hasn't traveled as far and so is propagating slower than the speed of light.

To show this, "that's a lot of math. So, you can do it by solving lots of equations...but it's a mathematical mess to get to that point."

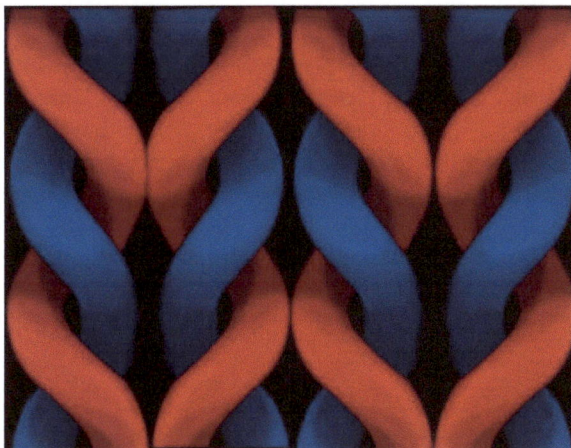

You can combine waves in a way to create a wave that travels faster then c, we are told. There are two speeds, phase speed and group speed. Group speed is always c, the phase speed can be faster than light."

He then relates this to the slit experiment, for an explanation according to QM. QM has one photon going through two slits at once. So the incident light is a superposition of the original light and all the other "interactions" that took place along the many different paths the photon took, the net result being a result of the photon interfering with itself. The net effect, he says, is slower moving light on the other side of the slit.

The interviewer asks a great question, "If the math works for both of them (the classical and quantum descriptions for this phenomenon), which one is actually happening?"

The answer is telling. "That's the problem with physics." This is physics, says he, where we can make many models of reality. "The quantum model is more right."

Another telling thing he says is that light in a vacuum is different than light in a lattice, and so light in a vacuum can explain the photon, but light and the lattice together, "maybe we can come up with a completely different particle." So the mathemagician used math to come up with something called the poleriton which is a combination of oscillations of the photon and the oscillations of "all the stuff in the material" the light is going through.

Awesome! When a particle interacts with other particles in a wave we end up with a new particle! Typical reification by theorists. They describe phenomena and think they are explaining objects!

The answer to "Why does light appear to be speeding up and slowing down?" is because of calculations of speed being made based on angles of incident light rather than actually measuring speed. More importantly, because they are holding frequency constant allowing wavelength and c to vary.

Of course, they are comparing a "normal path" with an angle of incidence at the boundary and this is said to be a ratio of the speed of light to speed through the material (called the refractive index).

So the speed of light (through a vacuum, and essential air) is constant, the only thing left is the angle. That's why they measure angles to calculate c. They measure frequency as a function of time (cycles per second), not length. Since they measure frequency as a function of time, rather than a function of numbers of links per unit length, the

ACCUMULATED variances in angles result in variances in computed time traveled.

Theorists confuse torsion with photons, and also believe that these particles travel from point A to point B rather than torsion being felt in both directions along the rope simultaneously. But torsion signals travel along many adjacent ropes extending from light source through the boundary on both sides of the material and beyond. They confuse this as wavefronts interacting. Torsion signals along one rope between any two atoms cause atoms in the material (lattice, as Dr. Moriarity says) to pump at a similar rate, and this results is more ropes torquing side by side. There are slightly different angles, and they add all these angles together to calculate light speed through different material.

They start with imaginary points on the aperture in the slit experiment, or on one boundary of the refractive material. These points become waves in a wavefront. The wavefronts interact with other particles forming more wavefronts and these all interact with each other as they travel along a path through to the other side. This is exactly what we are seeing with diffraction. Points becoming waves which interfere with each other.

So, what then is reflection?

Wiki says: "Reflection is the change in direction of a wavefront at an interface between two different media so that the wavefront returns into the medium from which it originated."

Just another angle, but one headed back in the direction it came. When light goes through the material, it refracts and reflects! So reflection, refraction, and diffraction are all torsion signals along different ropes than from the original source of torsion. In short, it is the accumulation of different angles which accounts for the calculation of a slower composite wavefront exiting the material. Mathematically the quantum mechanic says that frequency has to do with time and wavelength has to do with length (as the name implies). They hold frequency the same and so wavelength and c are reduced. c = f x wavelength.

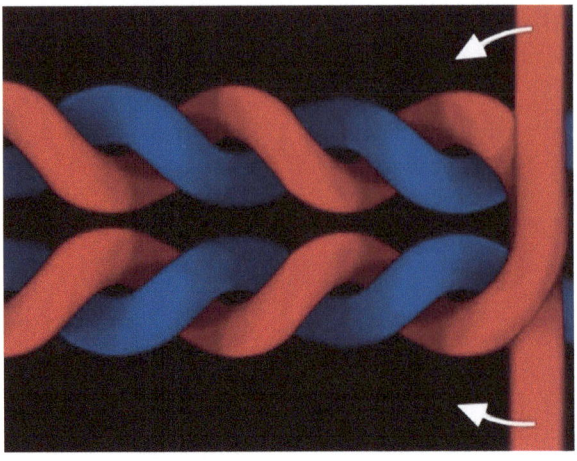

In a previous discussion David and I had, I said this, "Frequency is changing, yes, but amplitude must also be changing." BUT, there is no direct connection between frequency and amplitude on a single rope. Just like in the fringes caused by diffraction where there is a bright center band caused by many more ropes torquing in unison, in the example dealing with sound, many more ropes transferring sound are torquing in unison (harmonics, a kind of constructive interference) causing me to hear a louder tone.

Wolfram says reflection is "the phenomenon of a propagating wave (light or sound) being thrown back from a surface." But, what are they talking about? In typical fashion, the theorist is using math to describe the phenomenon without shedding any light on the subject of its cause, at all.

"In physics, a wavefront is the locus of points characterized by propagation of position of the same phase: a propagation of a line in 1D, a curve in 2D or a surface for a wave in 3D." – Wiki

The mathematical theorist ascribes points along a surface, such as on a mirror, and thinks these assumed point sources are particles (in a wavefront) which form secondary wavefronts that interfere with each other as they bounce back and forth between the mirrors.

Of course, a wave does not have a surface, as a wave is a location in which one is supposed to find particles.

According to Rope Hypothesis, light is a two way mechanism. All atoms are interconnected by a helically wound two strand rope that acts as a highway for torsion signals (light). An atom pumps, creating friction along its outer surface which is "felt" along with torque which "travels" both directions simultaneously along the ropes between any given two atoms.

Atoms in a light source, a reflective surface, the object being reflected, and an observer are all interconnected. Let's look at this and see if we can understand what is happening. When the light source is turned on, or is directed at an object, it causes the atoms in the object to pump and this causes torsion which is relayed along the ropes between atoms. The atoms in the light are connected to the atoms in the eyes of the observer. Regardless of whether the light is on or off, atoms are pumping and torsion signals are "felt" along all ropes.

Atoms never stop pumping. They just pump differently (amplitude and frequency) in the light's filament when energized by its power supply. The rope is comprised of an electric and magnetic thread which is wrapped around each other forming links. Frequency is the number of links per unit length, and amplitude is the height of each link. When the light is turned off, frequency shifts to one that is not perceived by human eyes.

Now forget about the light source except to realize that in order for an observer to see something, the torsion signals relayed to the eyes must be arriving along ropes with a number of links that have a frequency falling within our visible range. IOW, whether we are there, are paying attention, or have eyes, atoms are torquing signals in both directions along ropes between atoms.

A reflection just means that torsion signals are being relayed to an observers eyes rather than arriving directly from the source. If someone shines their flashlight on the wall, you can see it from the side because of light being reflected through atoms in the air.

Static Electricity and Thread Theory

What is static electricity? Seems dynamic to me!

Mom, Dad, and I were on our way to dad's job at the Air Force base in Tucson, Arizona. Mom needed the car that day and I tagged along. Dad was driving and I was in the back seat.

The funniest and oddest thing happened that day.

Dad kept an old wool army blanket over the front seat to protect it from wear. It was a cool, dry morning in the desert. A perfect day for what was about to take place. As Dad got out of the car Mom slid over into the driver's seat. Dad leaned over and stuck his head through the open driver's window to kiss Mom goodbye. Even before their lips touched, a blue spark jumped across to connect them in a spark of love. The static discharge not only zapped them both, causing them to jerk their heads back, it leaped across the gap between Dad's chest and his top fatigue pocket where it ignited a kitchen match!

The cool desert air, a wool blanket, and two unwilling participants, perfect for a static discharge. A static discharge in perfect conditions can produce an electrical voltage in excess of 25 thousand volts!

The mainstream explanation for static electricity might go something like this:

An atom is the smallest unit of matter that has the properties of a chemical element, such as, hydrogen, oxygen or helium, etc.

Consider the planetary model of the atom with a central nucleus comprised of protons and neutrons. Orbiting around the nucleus are electrons. Of course, this is just a useful way of looking at it because actually, you see, the electrons would be moving around the nucleus in various 3 dimensional configurations called orbitals.

A property intrinsic to these sub atomic particles, that is, protons, neutrons and electrons, is charge. Protons are positive, electrons are negative and neutrons are neutral. IOW, protons have a cheery outlook on life, electrons are always down and neutrons just don't give a damn!

The power or strength of the charges is equal, so each electron and proton gets one vote. If there are an equal number of Es and Ps, the atom is neutral, not wanting to commit to either side of the argument. The atom would prefer not to cause any static. There's enough trouble in the world as it is!

Sometimes electrons are not content to stay in the atom and will move to another atom. The atom it leaves will then be more positive because it didn't like the electron anyways. The atom that receives the new electron will be more negative because, well, there goes the neighborhood!

The two newly formed alliances are charged, now, and we call then ions.

Some matter holds on to its electrons very tightly. They are a tightly knit community called insulators. Other types of matter are loose with their electrons and they tend to move from atom to atom very readily. We call these conductors. What sort of business they are conducting, we don't really know (but it's none of OUR business anyways).

Positive objects tend to be drawn to negative ones for some reason, and they like to hang together. Well, you've heard the saying opposites attract. It must be true. Personally I believe birds of the feather flock together, but that's just me. A positively or negatively charged object will attract something that is neutral. The neutral object wants to hang with either because it can hang around any crowd, being non-committal and all.

Even though the object itself is attracted, if it is a good conductor, the electrons are reluctant and move away from the charged object towards the other side. On the other hand, if it is a good insulator, the electrons don't move as far away. In either case, the two objects will stick together until electrons from the charged object get tired of hanging around so many positives and they leave making an excuse that they have to take a leak.

Static electricity builds up as electrons migrate in this manner from one object to another. Then some dumb ape comes along and gets the shit zapped out of him. It's too funny! His fur stands up on end, matches catch fire and expletives abound.

BUT where do those electrons really go that said they had to take a leak? Who knows? Who cares? We are better off without them.

When two objects rub against each other, how do we know which one will become more negative and which more positive? It's based on a pecking order we call the triboelectric series.

No new electrons or protons were created or destroyed, they just moved somewhere else. The silly ape that got zapped gets a few more electrons, too, because of the conservation of charge principle. It just that he was higher up on the triboelectric series of objects. AND this dumb ape wants to become man moving higher up the phylogenetic tree. If you think it is funny watching an ape get shocked, you should see what happens when a man gets zapped!

Of course, that's just a bedtime story for your entertainment. Later on I'll fill you in on what is probably really happening using the rope model of the atom.

Here is a static discharge device to take the shock for you on "staticy" days:

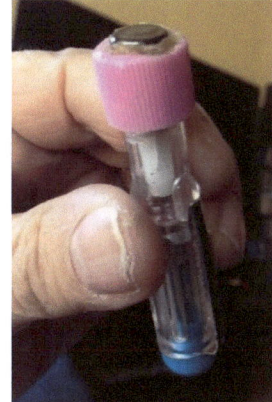

Here is another that I made years before those others were available.

Mine is just a modified screwdriver/circuit tester. Theirs is a smiley face that turns to a frowny face when it gets zapped.

How do these work, that is, based on the fairy tale above?

Mine is grounded through me when I hold it touching the clip. The current "flows" through the contact "into" the light filament and "to" the clip, before "traveling through me into the earth."

Theirs has an insulator (black rubber end) and you touch the silver button. This works essentially the same except they have a neat little circuit that turns a smile into a frown.

Keep in mind the triboelectric series.

This reminds me of something we used to do as kids on my Aunt's farm.

She had an electric fence to keep the cows and horses in the field. My brother and two cousins would walk along the fence next to each other. The bravest one closest to the fence would randomly grab the wire in one hand and grab the wrist of someone next to him. In anticipated response, that person would grab the closest person and so on down the line. Of course, none us of were faster than electricity, but if this was done "right," the person at the end of the chain received all the shock. That is, that person felt the shock the most.

Let's start over and see if we can offer RATIONAL explanations for this phenomenon.

Mainstream: An atom is the smallest unit of matter that has the properties of a chemical element, such as, hydrogen, oxygen or helium, etc.

Rational Explanation: An atom is the smallest unit of matter as typically defined; the set of all atoms. These atoms are hydrogen atoms which form all other atoms or elements, molecules and objects.

Mainstream: Consider the planetary model of the atom with a central nucleus comprised of protons and neutrons. Orbiting around the nucleus are electrons. Of course, this is just a useful way of looking at it because actually, you see, the electrons would be moving around the nucleus in various 3 dimensional configurations called orbitals.

Rational Explanation: Rope Hypothesis assumes a single closed loop thread which forms a two strand electromagnetic rope connecting all atoms. The M and E threads are helically wound, forming links and separating at the atom. The E thread continues through the center of the atom joining others and forming a proton dandelion, then continuing out the other side. The M thread wraps around the atom along with the other M threads forming the electron shell and then joins the E thread and continues on to another atom. The rope reels in and out as the atom pumps, or quantum jumps, torquing the ropes. The E shell is comprised of layers of threads that can rotate or revolve around the proton confusing the Phiz Whiz who concocts orbitals, clouds or probability distributions.

Mainstream: A property intrinsic to these sub atomic particles, that is, protons, neutrons and electrons, is charge. Protons are positive, electrons are negative and neutrons are neutral. IOW, protons have a cheery outlook on life, electrons are always down and neutrons just don't give a damn!

Rational Explanation: The protons and electrons act independently of each other vibrating, rotating, and revolving. The electron revolves in the opposite direction of the proton due to the pumping atom and torsional action of the rope. The neutron remains relatively stationary in relation to the atom.

Mainstream: The power or strength of the charges is equal, so each electron and proton gets one vote. If there are an equal number of Es and Ps, the atom is neutral, not wanting to commit to either side of the argument.

The atom would prefer not to cause any static. There's enough trouble in the world as it is!

Rational Explanation: Charge is a property attributed to the rotational orientation, hence neutrons are neutral.

Mainstream: Sometimes electrons are not content to stay in the atom and will move to another atom. The atom it leaves will then be more positive because it didn't like the electron anyways. The atom that receives the new electron will be more negative because, well, there goes the neighborhood!

Rational Explanation: Electron shells, or layers of the shell, can extend out at a greater distance than typical for the atom and does this for various reasons, such as

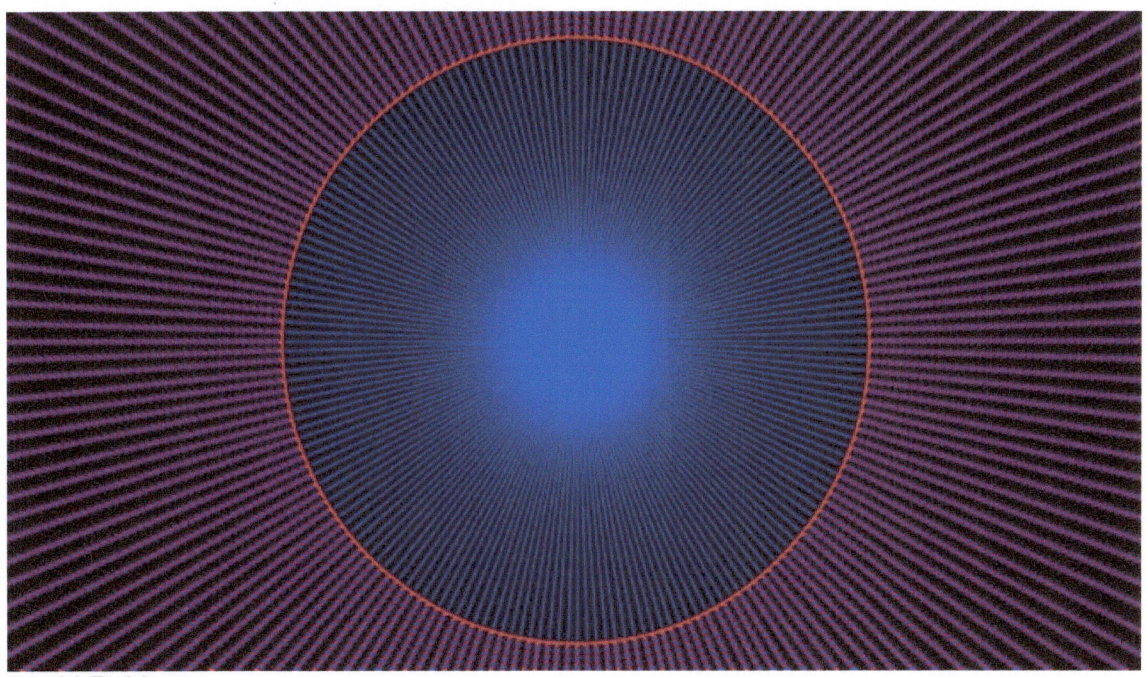

David Robison

an excited state where many nearby atoms are pumping and torquing at a higher rate. When an E shell, in its excited state, touches the E shell of a neighboring atom, this interaction is responsible for what the Whiz is saying about atoms being MORE positive or more negative.

Mainstream: The two newly formed alliances are charged, now, and we call then ions.

The atom, at this point, is called an ion. Some matter holds on to its electrons very tightly. They are a tightly knit community called insulators. Other types of matter are loose with their electrons and they tend to move from atom to atom very readily. We call these conductors. What sort of business they are conducting, we don't really know (but it's none of OUR business anyways).

Rational Explanation: Insulators have E shells that don't readily get excited out of their natural state. Conductors, on the other hand, have E shells that easily expand away from the proton and interact with neighboring atoms. This is due to the atomic arrangement in various materials. Although both are comprised of carbon, graphite is a good conductor of electricity but the diamond is not. It is the way the atoms are arranged differently in these materials that accounts for this. It is related to how the atoms are bonded. For now, let's just say that less of the E shells with the diamond are available for what we call conduction.

Mainstream: Positive objects tend to be drawn to negative ones for some reason, and they like to hang together. Well, you've heard the saying opposites attract. It must be true. Personally I believe birds of the feather flock together, but that's just me.

Rational Explanation: Objects, whose E shells are spinning mostly in opposite directions from another object's, tend to attract.

Mainstream: A positively or negatively charged object will attract something that is neutral. The neutral object wants to hang with either because it can hang around any crowd, being non-committal and all.

Rational Explanation: Whenever an object with many E shells rotating mostly in one direction comes into close proximity of another object which is neutral (E shells spinning mostly in all different directions), it will attract that object.

Mainstream: Even though the object itself is attracted, if it is a good conductor, the electrons are reluctant and move away from the charged object towards the other side. On the other hand, if it is a good insulator, the electrons don't move as far away. In either case, the two objects will stick together until electrons from the charged object get tired of hanging around so many positives and they leave making an excuse that they have to take a leak.

Rational Explanation: A good conductor is one whose E shells expand out away from the proton and interact with other E shells easily where insulator's E shells do not. Therefore, it makes sense that these objects interact the way they do.

Mainstream: Static electricity builds up as electrons migrate in this manner from one object to another. Then some dumb ape comes along and gets the shit zapped out of him. It's too funny! His fur stands up on end, matches catch fire and expletives abound.

Rational Explanation: When enough E shells are interacting in the manner described, it manifests in what we call a static discharge, or electrostatics.

Mainstream: BUT where do those electrons really go that said they had to take a leak?

Rational Explanation: Electrons or E shells don't go anywhere; E shells expand and M threads interact with other atom's M threads, for a time, and then return to normal. Think of the domino effect. A domino falls against another one knocking it down and then it returns to its "rest state."

Mainstream: When two objects rub against each other, how do we know which one will become more negative and which more positive? It's based on a pecking order we call the triboelectric series.

No new electrons or protons were created or destroyed, they just moved somewhere else. The silly ape that got zapped gets a few more electrons, too, because of the conservation of charge principle. It just that he was higher up on the triboelectric series of objects. AND this dumb ape wants to become man moving higher up the phylogenetic tree. If you think it is funny watching an ape get shocked, you should see what happens when a man gets zapped!

Rational Explanation. The higher up in the triboelectric series, the better the conductor. Better conductor's have atoms arranged so that their E shells more readily can expand and interact with neighboring E shells. Now something needs to be mentioned ....Charge. "Charge is composite friction around the surface of the electron during quantum jump." - WGDE, p.202, Fig. 433/

So, negative and positive charge is spin direction and elementary charge is relayed composite friction at the molecular level. We have to look at Coulomb's Law to get a better grasp of this. Spin is NOT a quantity, it is a quality. E shells never stop spinning, so charge "is conserved."

What is being measured is spin, a single link of rope at the surface of an atom as it is taken in. This measurement is interpreted as a quantity of electrons moving past a single point. An ion is an E-shell (or layer of M threads) of an atom that is/are extending out away from the atom.

Charge is the combined effect of atom's spinning E shells and their interaction with other E shells (or group of atoms and their E shells - molecules)

Negative and positive charge just refers to direction of one E shell's spin against another, or group of atom's and their E shells (molecules).

Elementary charge is the direction of spin of an electron or proton (negative or positive). ACTUALLY, spin direction of protons probably does not even figure into this. E and M threads arrive at the surface of the atom and split off, then "reunite" at the opposite side of the atom. So not only are there layers of M threads spinning and rotating in all different directions at the E shell, there are E threads spinning and rotating in all different directions in the proton. However, the M and E threads are PAIRED. Therefore, as far as that pair is concerned, the "electron" and "proton" are moving in the same direction, but not necessarily in relation to other EM pairs. It is at the rope where E an M threads are spinning in opposite directions as they unravel at the atom.

Repulsion/attraction is the result of M-threads interacting with other M- threads.

I assume that while electrical and magnetic "forces" are infinitesimally small at thread, rope and atom level, at a molecular level 25 thousand volts can be significant. Just ask my Dad or the guy that pisses on an electric fence.

This phenomenon of charge, then, is the relation between fundamental and composite objects!

So why is static stronger on cool dry days?

Come by the Rational Scientific Method or Rope Hypothesis facebook pages for a discussion with the luxury of detail.

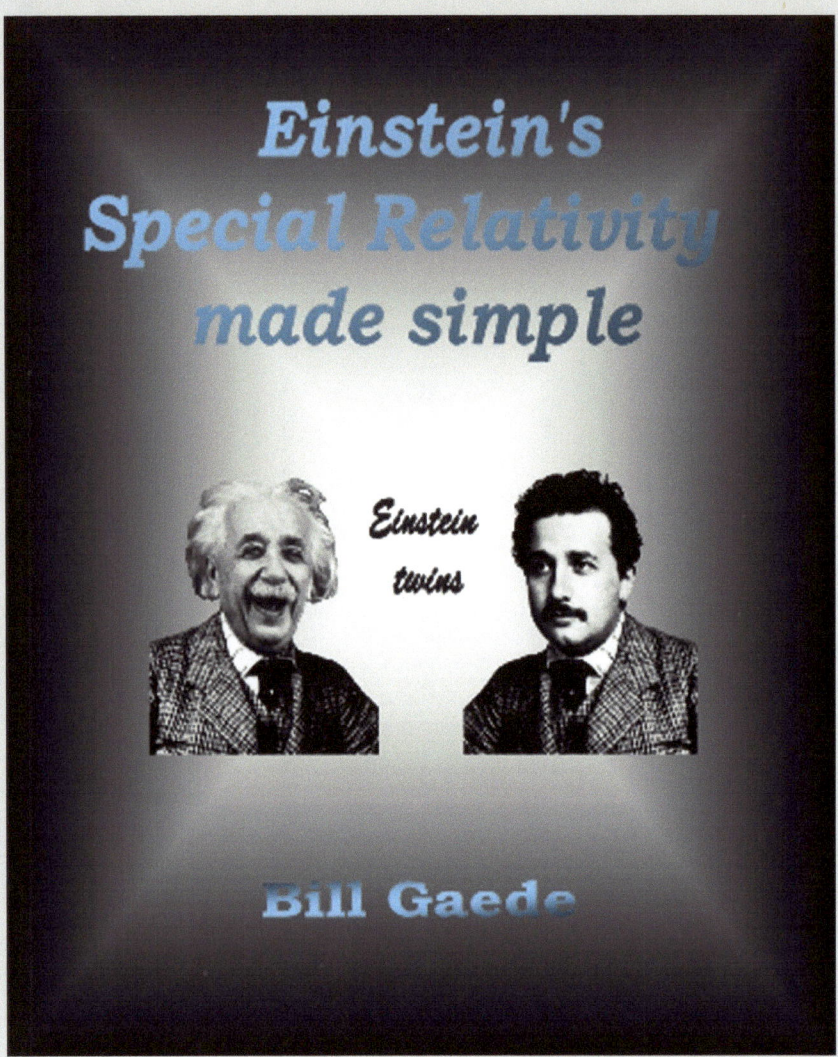

Special Relativity made simple is the ideal book for laymen with open minds who suspect the fantastic explanations they hear and read about from authoritative 'scientific' sources. Does it make sense to say that twin brothers could differ in their ages by 50 years? Is it rational to say that you can travel to the past or to the future?

Special Relativity made simple addresses these and similar questions and arms the average reader with arguments that enables them to challenge Einstein's theories on the Internet and in conferences.

To obtain a paperback, Paypal to bill@youstupidrelativist.com
USA/Canada   US $20.00 Plus $10 Shipping
Europe 20.00 Euros Plus 10 Euros Shipping

www.ingramcontent.com/pod-product-compliance
Lightning Source LLC
Chambersburg PA
CBHW040416220526
45473CB00004B/1262